CONTENTS

EXECUTIVE SUMMARY		i
CHAPTER ONE: BACKGROUND AND SURVEY METHODS		3
1.1	BACKGROUND	3
1.2	RESEARCH AIMS	4
1.3	METHODOLOGY: SURVEY APPROACH	4
1.4	SAMPLING	5
1.5	QUESTIONNAIRE DESIGN	6
1.6	RESPONSE	6
1.7	CHARACTERISTICS OF SURVEY RESPONDENTS	7
CHAPTER TWO: SURVEY RESULTS		9
2.1	INTRODUCTION	9
2.2	LIKES AND DISLIKES ABOUT THE WIND FARM	9
2.3	VISIBILITY OF WIND FARMS	11
2.4	ANTICIPATED AND ACTUAL PROBLEMS WITH THE WIND FARMS	13
2.5	ANTICIPATED AND ACTUAL BENEFITS FROM THE LOCAL WIND FARM	16
2.6	SOURCES OF INFORMATION AND COMMUNICATION	19
2.7	ATTITUDES TOWARDS POSSIBLE FUTURE DEVELOPMENT	21
2.8	LEVELS OF ENVIRONMENTAL CONCERN	24
CHAPTER THREE: CONCLUSIONS		27
APPENDIX ONE: BACKGROUND INFORMATION ABOUT THE WIND FARMS		29
APPENDIX TWO: THE QUESTIONNAIRE		33

EXECUTIVE SUMMARY

Background and introduction

System Three Social Research was commissioned by the Scottish Executive to undertake research examining the attitudes of local populations towards the four operational wind farms in Scotland (Hagshaw Hill, South Lanarkshire; Windy Standard, Dumfries & Galloway; Novar, Highland and Beinn Glas, Argyll & Bute). The major aim of the research was to examine how local residents feel about the existence and proximity of their local wind farm. An important objective was to identify whether, and to what extent residents' views of wind farms are based on actual experience or perception formed through the media, word of mouth or other sources.

The sample for the survey was drawn from all residents living within 20km of a wind farm in Scotland and the number of interviews were distributed across three different proximity zones within the 20km radius. Residents within 5km of a wind farm were defined as being in the high proximity zone, residents between 5km and 10km in the medium proximity zone and residents between 10km and 20km from a wind farm in the low proximity zone.

Overall, 430 telephone interviews were conducted over the three proximity zones. The distribution was deliberately skewed towards residents in the high proximity zone where 215 interviews were conducted. The remainder of the interviews were evenly split between the medium and low proximity zones.

Survey results

Respondents were generally positive about wind farms. Those who lived nearest a wind farm were more likely to provide positive responses when asked about the wind farm than those in the other zones. For example, while 67% of respondents overall said that there was something they liked about the wind farm, this proportion increased to 73% of those living in the high proximity zone.

Overall, 74% of respondents said there was nothing they *disliked* about the wind farm. Only 11% said that there was nothing they *liked*.

The proportion of respondents who had anticipated problems prior to the development (40%) was far higher than the proportion who actually experienced problems after the development (9%). Actual noise caused by the turbines or the visual impact of the wind farm did not feature as issues for the majority of respondents.

Although 12% of all respondents had expected to experience a problem with noise, only 1% of respondents had actually experienced a problem and only 2% said unprompted, that they disliked the wind farm because it was noisy.

In relation to the visual impact of the wind farm, a higher proportion of respondents said that they liked the look of the wind farm (21%) than the proportion who said the farm was unsightly or spoiled the view (10%).

Respondents anticipated and experienced benefits from the wind farms. The proportion of respondents who anticipated benefits was higher than the proportion who actually experienced them. Overall, 63% anticipated at least one benefit, while 32% said they had experienced a benefit. However, many residents had experienced benefits. For example, 18% said that local jobs had been created and this proportion roses to 26% in the high proximity zone. Similarly, 7% overall and 11% of respondents in the high proximity zone said that there had been an increase in tourism.

Respondents indicated low levels of awareness of consultation either by the wind farm developer or the local authority. Only 20% were aware of consultation being conducted by the wind farm developer and 17% were aware of consultation by the local authority. The most common source of information about the wind farm was from local newspapers (42%), with only 7% gaining information from the local authority or the developer.

When asked about potential future wind farm developments, the positive attitude towards wind farms was reflected in the fact that only 14% of respondents would be concerned if extra turbines were added to the existing wind farm.

Twenty seven percent would be concerned if another wind farm was proposed for the local area. Respondents who can see the wind farm from their homes, are more likely to be concerned about another wind farm proposal than those who see the farm in other circumstances.

Although respondents were generally positive about the farms and most were not concerned about future development, the majority thought that wind farms should be located in uninhabited areas and high on hills.

The majority of all respondents currently living near wind farms had not experienced any problems with the wind farms. The problems they had anticipated did not materialise in the vast majority of cases (only 9% of respondents experienced problems compared with 40% who anticipated them). This suggests that the information provided about wind farms and the explanations for their development are crucial in reducing this anticipation of problems.

CHAPTER ONE: BACKGROUND AND SURVEY METHODS

1.1 BACKGROUND

1.1.1 System Three Social Research was commissioned by the Scottish Executive to undertake research examining the attitudes of local populations towards the four operational wind farms in Scotland (Hagshaw Hill, South Lanarkshire; Windy Standard, Dumfries & Galloway; Novar, Highland and Beinn Glas, Argyll & Bute). The research results are intended to inform the Scottish Executive's renewable energy policy, in particular the current review of National Planning Policy Guideline 6 (NPPG6): Renewable Energy (see NPPG 6 (Draft Revision June 2000) : Consultation Draft issued 1st June 2000).

1.1.2 The UK Government and the Scottish Executive are committed to increasing the proportion of electricity generated by renewable energy sources. In June 1999, the Government announced that it wanted 10% of the UK's electricity to be produced from renewable sources by 2010. This policy to increase the use of renewable energy will be promoted through the Renewables (Scotland) Obligation, which will replace the previous Scottish Renewables Obligation. It will require all electricity suppliers to source a specified proportion of the electricity they provide for their customers from renewables generation. Additionally, this policy is promoted through planning policy guidance (NPPG6) which encourages local authorities, among other considerations, 'to provide positively for renewable energy developments, where this can be achieved in an environmentally acceptable way' (NPPG 6 (1994), para.25). NPPG6 (1994) also identifies that Scotland has particular potential for the development of renewable energy, especially wind energy, due to its geography and climate.

1.1.3 Although Government support and suitable conditions for wind farms exist in Scotland, concern has been raised that several viable proposals over the past decade have failed to secure planning permission. In Scotland, only around two thirds of wind farm applications have been successful compared with 85% of renewable projects overall and 90% of all applications. These figures suggest that wind farm applications are more likely to fail than applications of any kind and, than other renewable energy projects.

1.1.4 Objections to wind farm developments have been based on a range of issues from visual intrusion to ecological effect. The 1994 NPPG 6 identifies a range of potential issues related to wind farms and outlines the considerations necessary to minimise their effect. The main potential issues outlined in NPPG6 are;

- Visual impact
- Noise from the turbines
- Interference, eg with telecommunications or driver distraction
- Environmental or ecological effect

1.1.5 NPPG6 states that, in addition to general planning considerations, the potential impact of these issues should be examined during the planning process for any wind farm development. For example, wind turbines should not be permitted where they would create noise problems for residential or other noise-sensitive properties.

1.1.6 Despite the considerations outlined in the planning guidelines, objections to wind farms at the planning stage have been based on the potential noise or visual effect of the proposed farm. As there is likely to be an increase in the number of planning applications for wind farms in Scotland, it is necessary to examine in more detail the actual experiences of people who live near existing wind farms in order to inform the planning process.

1.1.7 Wind farm proposals have tended to generate controversy, but research undertaken on behalf of the wind power industry shows that local residents tend not to be as opposed to wind farm developments as the number of planning refusals suggest. Independent research conducted in England and Wales suggests that local residents are largely in support of the nearby wind farms. Previous research conducted on behalf of the wind power industry in Scotland reached similar conclusions. However, as much of the previous research has been funded by the wind power industry, there is scope for accusations of partiality.

1.2 RESEARCH AIMS

1.2.1 In order to inform both developers and the relevant authorities, it was necessary for independent and methodologically robust research to be conducted to ascertain the true opinions and perceptions of local residents to wind farms. The research focused on a sample of residents living near the four existing wind farms in Scotland. These farms are, Beinn Glas in Argyll and Bute, Hagshaw Hill in South Lanarkshire, Novar in Highland and Windy Standard in Dumfries and Galloway.

1.2.2 The major aim of the research was to examine how local residents feel about the existence and proximity of their local wind farm and whether looking back, residents felt that their views of wind farms had changed.

1.2.3 Within this overall aim, an important objective was to identify whether, and to what extent, residents' initial views of wind farms had been based on actual experience, or perception based on the media, word of mouth or other information sources.

1.2.4 The report is structured as follows: Chapter Two outlines the methodology of the research, Chapter Three provides the survey results and Chapter Four concludes the report. Appendix One provides background information about the wind farms, public opposition or support during the planning process and any subsequent problems or complaints. Appendix Two contains a copy of the questionnaire.

1.3 METHODOLOGY: SURVEY APPROACH

1.3.1 Due to the geographically dispersed nature of the populations around the four Scottish wind farms, the method of telephone interviewing was selected as the most cost-effective approach.

1.3.2 In order to examine the local populations' attitudes to wind farms, it was first necessary to identify what constituted the local population. In other words, how near to one of the wind farms would it be necessary for someone to live before they would be included in the survey.

1.3.3 This was approached by examining the potential issues associated with living near a wind farm (for example, noise from the turbines or an adverse visual impact) and the likely distance that these would potentially have an effect. These considerations, in conjunction with findings from previous research undertaken by other organisations and consultation with wind farm operators and local authority planning personnel, was used in assessing suitable areas for inclusion in the study.

1.3.4 The visual effect of a wind farm or turbine is arguably the issue with the largest geographical range and obviously varies widely depending on the farm's situation. Equally, peoples' perceptions of the effect vary, with some believing the landscape is enhanced and others believing it is destroyed. Previous research[1] and consultation with planning authorities identified that the scope of the visual effect has, on occasion, been under-estimated and that the views of those up to 15km to 20km away can be affected. On the basis of this, it was decided that all residents within 20km radii of the wind farms should be included in the sample, which would allow inclusion of a larger proportion of people who, in their daily lives might see the wind farm. This also allowed for the inclusion of residents who perhaps can't see the turbines from their home but see it whilst, for example, driving to work. This also enables views to be examined on driver distraction, shadow flicker, disruption during construction etc.

1.3.5 However, residents living within the 20km radius, but at different proximities from the wind farms, are likely to have different experiences of the wind farm development. Taking 20km as the maximum range allows residents on whom the wind farm might only have had a minimal effect to be included. Residents living closer to the wind farms are likely to have more experience of the wind farm and potentially are better able to comment on the issues. Therefore, the 20km radius was divided into three zones of proximity to the wind farms – high proximity, medium proximity and low proximity.

1.3.6 The high proximity zone includes residents living up to 5km from a wind farm. It is likely that residents within these areas will be more concerned about the effects of the wind farm.

1.3.7 Residents living more than 5km away but less than 10km are in the medium proximity zone. Residents in this category may perhaps not be affected by noise, for example but are likely to be able to comment on the visual effect, disruption during construction, the relationship with the wind farm developer and the impact on the local economy and environment.

1.3.8 Residents living over 10km but less than 20km from the wind farms were included in the low proximity zone.

1.4 SAMPLING

1.4.1 The intention was not to provide separate analysis for each wind farm but to analyse the views of residents in each of the zones across the four wind farms. Therefore, three samples were drawn encompassing all residents who lived in the three zones of all four wind

---

[1] Chris Blandford Associates 1994: Wind Turbine Power Station Construction Monitoring Study, Countryside Commission for Wales.

farms. The sample was based on distance lived from any of the four wind farms rather than distance from a specific wind farm.

1.4.2 The first stage of the sampling involved obtaining the Ordnance Survey grid references of the four wind farms, mapping the three zones for each wind farm and fitting these zones to match with postcode geography. The addresses contained within the three postcode zones were obtained and a sampling organisation provided the telephone numbers for the addresses where available. The final sample was therefore drawn from residents within 20km of a wind farm who have a telephone. Around 95% of the total Scottish population have a telephone and this is slightly higher in rural areas. So the exclusion of people without a telephone does not bias the results.

1.4.3 The distribution of the total number of interviews across the three zones was deliberately skewed towards the high proximity zone, due to the importance attached to wind farm proximity. An overall target of 400 interviews was set; 200 with residents in the high proximity zone and 100 with residents in each of the other two zones. However a total of 430 interviews was achieved.

1.5 QUESTIONNAIRE DESIGN

1.5.1 As the survey aimed to examine residents' experiences of, and attitudes towards, living near wind farms in Scotland, many of the questions were left open in order to allow a free expression of opinion and experience, rather than suggest potential problems or benefits to respondents.

1.5.2 Additionally, certain specific issues commonly associated with wind farms (either factually or perceptually) were explicitly examined in the questionnaire to ascertain whether residents did experience problems with, for example, noise, the visual effect, or television or radio interference.

1.5.3 To examine whether respondents' views were shaped by actual experience or by perception (based for example, on the media, word of mouth or other information sources), respondents were asked what they thought might be the negative and positive aspects of the wind farm when they first learnt about its existence or its proposed development and then which of these they actually experienced.

1.5.4 General attitudes towards environmental issues and renewable energy were also examined to explore how this may impact on attitudes to wind farms. A copy of the questionnaire is included as Appendix Two.

1.6 RESPONSE

1.6.1 Overall, 430 interviews were conducted with individuals in households living within 20kms of one of the four wind farms. Of these, 50% (215) were with people within 5km of the wind farm, 25% (108) over 5km but less than 10km and 25% (107) over 10km but less than 20km.

1.6.2 The deliberate skewing of the sample towards the high proximity zone means that a high proportion of interviews were conducted with residents living near Novar, as this farm has the highest number of residents in close proximity. The views of residents living nearest wind farms are over-represented in the results because it is likely that they will have most experience of the wind farms.

1.7 CHARACTERISTICS OF SURVEY RESPONDENTS

1.7.1 The views of a range of different ages, household types and socio-economic groups are included in the survey. Overall, 40% of respondents were male and 60% female. 17% were aged 16 to 34, 40% aged 35 to 54 and the remaining 44% aged 55 or over. Thirty percent of survey respondents had children.

1.7.2 A high proportion of respondents (88%) had lived in the local area during the planning process and development of the wind farm. This is important as it enables analysis of residents' expectations when the farm was proposed alongside their actual experiences.

CHAPTER TWO: SURVEY RESULTS

2.1 INTRODUCTION

2.1.1 This section presents the findings of the survey. It examines what local residents like and dislike about the wind farm, how their actual experiences differ from their expectations when they first learnt about the wind farm. The section also considers how attitudes differ depending on proximity to the wind farm and how often and in what circumstances the wind farm is seen by residents. General attitudes towards the environment are also examined by use of an 'environmental indicator' which is constructed using respondents' answers to a set of questions relating to concern about the environment[2].

2.2 LIKES AND DISLIKES ABOUT THE WIND FARM

2.2.1 Respondents were first asked what, if anything, they like about having a wind farm in their local area. This was an open question which allowed respondents to spontaneously mention any aspect of the wind farm. The responses of all surveyed residents over the three zones are presented in Table 1.

Table 1: What residents like about the wind farm (%)

Likes	High proximity	Medium proximity	Low proximity	Overall
Not doing any harm/no bother	28	24	28	27
Landmark/looks good/interesting	25	27	7	21
Environmentally friendly	18	23	14	18
Nothing	10	10	13	11
Cheap electricity	6	11	3	6
Good for local economy	9	4	-	5
Better than fossil fuel	3	5	-	3
Better than nuclear power	3	3	3	3
Don't like it/bad idea	3	5	-	3
Other	5	2	11	6
Don't know	17	19	30	21
Base	215	108	107	430

Note for table
Nil per cent is indicated by a dash (-); figures between 0% and 0.5% are indicated by an asterisk (*); figures between 0.5% and 1% are rounded up to 1%. Percentages add up to more than 100 as respondents could give more than one answer.

---

[2] A positive response at each of questions 24-29 (see appendix two) was awarded one point creating an environmental indicator for each respondent on a scale between six and zero with six indicating the highest level of environmental concern and zero the lowest.

2.2.2 The most common response (27%) that the wind farm is 'not doing any harm' whilst not being overly positive, indicates that even if there is nothing that residents especially like about the wind farm, they do not mind its presence. Additionally, the fact that 21% said that they didn't know or hadn't thought about it suggests indifference rather than dislike.

2.2.3 More positively, 21% said that they actively liked the look of the wind farm or that it was a good landmark and 18% liked it because it was environmentally friendly or good for the environment.

2.2.4 There were only slight variations in responses from residents living in the different proximity zones. As might be expected, respondents in the low proximity zone were more likely to say that they had not thought about it. A slightly higher proportion of respondents in the high proximity zone said the wind farm was good for the local economy and that the wind farm looked good or interesting.

2.2.5 Although proximity to the wind farm is crucial, it is evident that some residents within the high proximity zone see the wind farm less often than some of those in the other two zones due to the local land form and location of the wind farm. Tables 4 and 5 on pages 11 and 12 show how often residents in the different zones see the wind farms and what proportions see them in different circumstances. It might have been expected that residents who see the wind farm very frequently or live in a high proximity area would be less positive about the wind farms. However, as can be seen from Tables 2 and 3, the opposite is the case.

2.2.6 Table 2 shows what residents who see the wind farms at different frequencies like about wind farms. In some of the categories the numbers are very small so the table only shows the 'likes' mentioned by 5% or more of respondents.

Table 2: What residents like about the wind farm by frequency of sighting (%)

Likes	Every/most days	Occasionally	Never	Overall
Not doing any harm/no bother	20	31	26	27
Landmark/looks good/interesting	32	19	5	21
Environmentally friendly	24	18	7	18
Cheap electricity	8	7	2	6
Good for local economy	8	4	4	5
Don't know	14	13	19	21
Base	133	240	57	430

2.2.7 As Table 2 shows, those who see the farm 'every' or 'most days' are more likely than those who see it 'occasionally' or 'never' to say that they like the look of the farm, or it is interesting or a landmark. Similarly, a higher proportion of residents who see the wind farm 'every day' say they like the farm because it is environmentally friendly or good for the local economy.

2.2.8 Residents who only see the farm 'occasionally' are more likely than those who see the farm more often to say that the farm is not doing any harm (31%).

2.2.9 Respondents were then asked what, if anything they particularly disliked about having a wind farm in the local area. Overall 74% said that there was nothing that they disliked and 8% said they did not know or had not thought about it. Table 3 shows the dislikes overall and across the three proximity zones.

Table 3: Residents' dislikes about the wind farm (%)

Dislikes	High proximity	Medium proximity	Low proximity	Overall
Nothing	80	72	64	**74**
Unsightly/spoils view	8	11	12	**10**
Noisy	1	-	5	**2**
Should be built elsewhere	2	-	1	**1**
Hazard to bird /wild life	-	1	2	**1**
Not necessary/ don't need it	-	2	1	**1**
Other	6	5	5	**6**
Don't know	4	10	13	**8**
Base	215	108	107	**430**

Note for table
Nil per cent is indicated by a dash (-); figures between 0% and 0.5% are indicated by an asterisk (*); figures between 0.5% and 1% are rounded up to 1%. Percentages add up to more than 100 as respondents could give more than one answer.

2.2.10 There was some variation across the proximity zones. Those living in the high proximity zone were more likely to say that there was nothing they disliked about living near a wind farm (80%) than residents in the other two zones. A slightly higher proportion of residents in the medium and low proximity zones said they disliked the wind farm because it was unsightly or spoiled the view, 11% and 12% respectively.

2.2.11 Only 1% (2 people) in the high proximity zone said they disliked the wind farm because it was noisy. Surprisingly, slightly more respondents in the low proximity zone said that they disliked the farm because of noise (although this was only 5 people).

2.2.12 None of the respondents mentioned the problem of 'shadow flicker' or 'driver distraction' when asked what they disliked about wind farms.

2.3 VISIBILITY OF WIND FARMS

2.3.1 As discussed above, although proximity is a factor which might influence attitudes towards wind farms, there are circumstances where people live near a farm but rarely see it because of the siting of the farm or the local land form. The circumstances under which residents see the wind farm could potentially influence how favourably they consider the farm. For example, a resident who can see the farm from their home or garden might be more concerned about the presence of the farm than someone who sees it while out hill walking. Similarly, how often respondents see the wind farms could perhaps influence their opinions. Table 4 shows the circumstances in which residents can see the wind farm. Overall, 13% can see a wind farm from their home or garden, 46% when travelling on local roads, 66% when travelling on major roads and 18% when in the country or hill walking. The table shows that those who live nearest wind farms are less likely to be able to see them from their homes and gardens than those living further away. A higher proportion of respondents living between 5km and 10km from a wind farm see it while travelling on local roads than those who live within 5km.

Table 4: Circumstances in which respondents see the wind farms (%)

	High proximity	Medium proximity	Low proximity	Overall
From home or garden	4	36	9	**13**
When travelling on local roads	41	68	30	**46**
When travelling on major roads	70	48	74	**66**
Out walking in the country/hill walking	15	29	11	**18**
On television	1	-	-	*
From everywhere/ all sides	-	1	-	*
Other	4	6	1	**4**
Don't know	-	1	2	**1**
Base	215	108	107	**430**

Note for table
Nil per cent is indicated by a dash (-); figures between 0% and 0.5% are indicated by an asterisk (*); figures between 0.5% and 1% are rounded up to 1%. Percentages add up to more than 100 as respondents could give more than one answer.

2.3.2   Table 5 shows how often respondents living in different proximity zones see the wind farm. Overall, only 15% see the farm 'every day', 16% see it 'most days', 56% 'occasionally' and 13% 'never'. A higher proportion of those living in the medium proximity zone (33%) see the wind farm daily than those living in the high proximity zone (11%). Similarly, a lower proportion of those in the medium proximity zone see the wind farm 'occasionally' than those in the high proximity zone.

Table 5: Frequency of circumstances in which respondents see the wind farms (%)

Frequency of sighting	High proximity	Medium proximity	Low proximity	Overall
Every day	11	33	7	**15**
Most days	17	19	8	**16**
Occasionally	63	35	62	**56**
Never	9	12	23	**13**
Base = 100%	215	108	107	**430**

Note for table
Nil per cent is indicated by a dash (-); figures between 0% and 0.5% are indicated by an asterisk (*); figures between 0.5% and 1% are rounded up to 1%.

2.4 ANTICIPATED AND ACTUAL PROBLEMS WITH THE WIND FARMS

2.4.1 To examine whether respondents' attitudes towards wind farms had changed, respondents were asked whether they had been concerned about certain specific issues when they first heard about the wind farm and then about actual problems they had experienced. Respondents were asked if they had anticipated problems with any of the following issues;

- Noise from the turbines
- The look of the landscape being spoiled
- Interference with TV and radio reception
- Damage to plants or animals
- Noise or disturbance during construction
- Extra traffic during construction
- A reduction in house prices.

2.4.2 Overall, 60% said that they had not anticicpated any of the listed problems, with this proportion increasing to 68% in the low proximity zone and decreasing to 57% in both the high and medium proximity zones. Nineteen percent of respondents were only concerned about one of the potential problems listed.

2.4.3 Table 6 shows the percentage of respondents overall and across the three zones who had been concerned about each issue. The most common concern overall had been about the look of the landscape being spoiled, which was mentioned by 27% of residents overall. Potential noise from the wind farm turbines had been a cause for concern for 12% of all respondents. A higher proportion of those who now see the wind farm 'every day' had been concerned about potential noise than those who saw it less often.

Table 6: Potential problems anticipated by respondents (%)

Anticipated problems	High proximity	Medium proximity	Low proximity	Overall
None	68	57	57	**60**
Noise from the turbines	13	13	9	**12**
Landscape being spoilt	29	28	26	**27**
Interference with TV/radio	7	12	5	**8**
Traffic during construction	10	16	2	**10**
Disturbance during construction	7	14	4	**8**
Reduction in house prices	6	6	7	**6**
Damage to plants/animals	14	16	7	**13**
Don't know	1	2	10	**3**
Base	215	108	107	**430**

Note for table
Nil per cent is indicated by a dash (-); figures between 0% and 0.5% are indicated by an asterisk (*); figures between 0.5% and 1% are rounded up to 1%.

2.4.4 The potential for the landscape to be spoiled had been a concern for slightly higher proportions of those in the high and medium proximity zones than in the low proximity zone.

Slightly fewer of those who would see the wind farm 'every day' had been concerned about the look of the landscape being spoiled than those who would see it 'most days'. Higher proportions of those living in the high and medium proximity zones anticipated problems with TV reception and traffic during construction than those in the low proximity zone.

2.4.5 Residents were asked an open question about whether they had had any other concerns when they learnt about the wind farm. Overall, 82% said that they had had no other concerns, 6% said they had never thought about it and 10% mentioned issues that have already been discussed. One percent said they had been worried that the farm might attract visitors and 1% had been worried about a negative effect on employment.

2.4.6 Respondents were then asked what problems they had actually experienced with the wind farm. Table 7 shows the problems actually experienced by respondents overall and across the three proximity zones. Overall, 87% said that they had not experienced any problems with the local wind farm. A further 4% said they did not know of any problems which means that over 90% of respondents have experienced no problems with the wind farm. Six percent had a problem with one issue and 5% had experienced problems with two or more issues. A slightly higher proportion of respondents in the high and medium proximity zones had experienced problems than those in the low proximity zone.

2.4.7 Only 2% of all respondents had experienced any problems with TV or radio interference.

Table 7: Problems experienced by respondents (%)

Actual problems	High proximity	Medium proximity	Low proximity	Overall
None	88	83	87	87
Noise from the turbines	1	2	1	1
Landscape being spoilt	6	6	3	5
Interference with TV/radio	1	5	5	2
Traffic/disturbance during construction	4	4	-	5
Damage to plants/animals	4	2	-	2
Don't know	1	6	9	4
Base	215	108	107	430

Nil per cent is indicated by a dash (-); figures between 0% and 0.5% are indicated by an asterisk (*); figures between 0.5% and 1% are rounded up to 1%. Figures may add up to more than 100% as respondents could give more than one answer.

2.4.8 Figure 1 shows the proportions of respondents that *actually experienced* problems alongside the proportions that *anticipated* problems. It is evident that the number of people who anticipated problems far exceeds the number of people who actually experienced them.

Figure 1: Anticipated and actual problems Base = 430

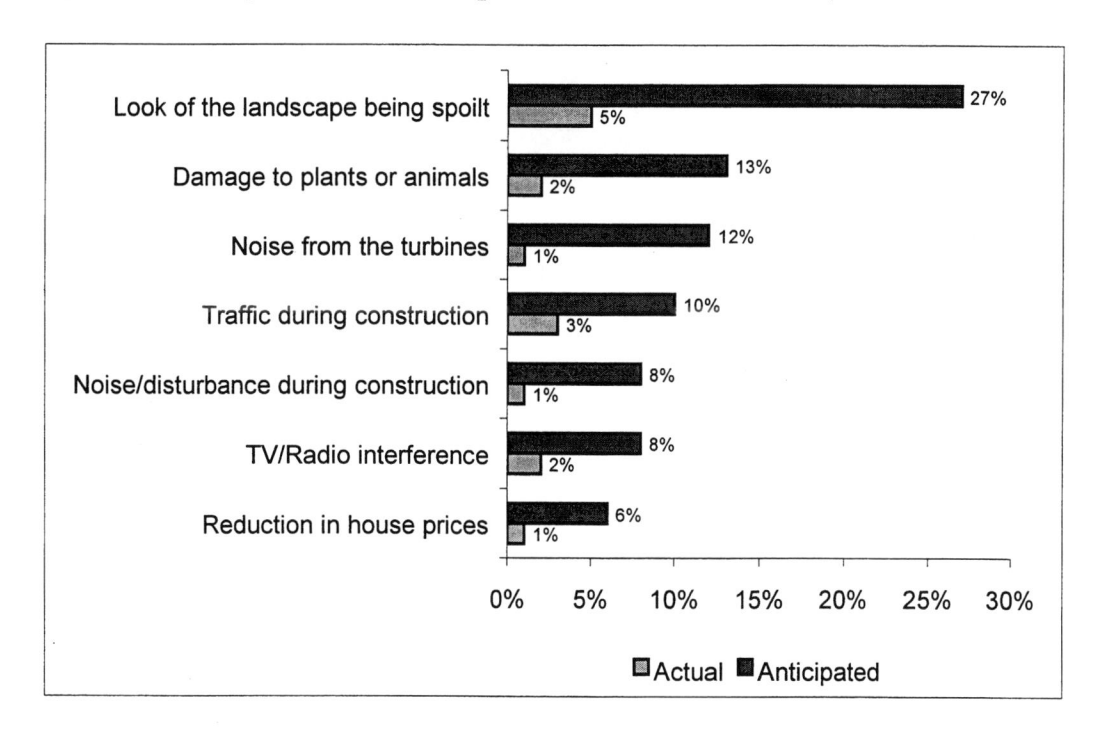

Problem	Actual	Anticipated
Look of the landscape being spoilt	5%	27%
Damage to plants or animals	2%	13%
Noise from the turbines	1%	12%
Traffic during construction	3%	10%
Noise/disturbance during construction	1%	8%
TV/Radio interference	2%	8%
Reduction in house prices	1%	6%

2.4.9 Figure 2 shows the same comparison but for the high proximity zone where it might be expected that respondents experienced more problems. However, as the figure shows, the difference between anticipated and actual experience shows a similar pattern to the overall figure, with anticipation of problems greatly outnumbering actual problems.

Figure 2: Anticipated and actual problems by residents in high proximity zone

Base = 215

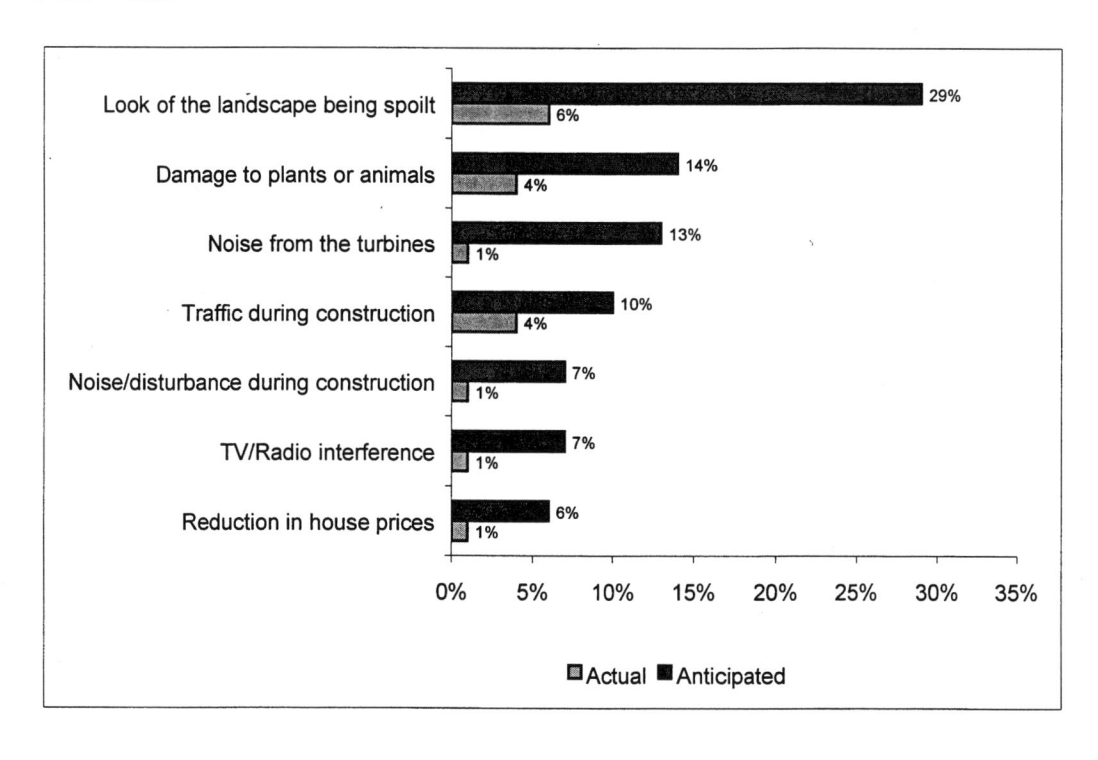

Problem	Actual	Anticipated
Look of the landscape being spoilt	6%	29%
Damage to plants or animals	4%	14%
Noise from the turbines	1%	13%
Traffic during construction	4%	10%
Noise/disturbance during construction	1%	7%
TV/Radio interference	1%	7%
Reduction in house prices	1%	6%

2.5 ANTICIPATED AND ACTUAL BENEFITS FROM THE LOCAL WIND FARM

2.5.1 Respondents were asked which of the following potential benefits they had anticipated when they first heard about the wind farm:

- Reduced pollution
- Cheaper electricity
- Electricity supplied locally
- Local jobs created
- Increased tourism

Table 8 shows the benefits anticipated by respondents overall and across the proximity zones.

Table 8: Benefits anticipated by respondents (%)

Anticiapted benefits	High proximity	Medium proximity	Low proximity	Overall
Cheaper electricity	50	34	30	**41**
Local jobs created	47	36	13	**36**
None	21	31	38	**28**
Electricity supplied locally	28	23	13	**23**
Reduced pollution	23	16	24	**22**
Increased tourism	12	12	1	**9**
Other	23	10	7	**16**
Don't know	5	9	16	**9**
Base	215	108	107	**430**

Note for table
Nil per cent is indicated by a dash (-); figures between 0% and 0.5% are indicated by an asterisk (*); figures between 0.5% and 1% are rounded up to 1%. Figures may add up to more than 100% as respondents could give more than one answer.

2.5.2 Overall, 63% said they had anticipated at least one of the listed benefits. This proportion rose to 74% of respondents in the high proximity zone and fell to 60% in the medium proximity zone and 46% in the low proximity zone.

2.5.3 Residents in the high proximity zone were more likely than those in the other two zones to have anticipated cheaper electricity (50%), electricity to be supplied locally (28%) and the creation of local jobs (47%).

2.5.4 Respondents were asked if there had been any other anticipated benefits from having the wind farm in their local area which they had not already been asked about. Sixteen per cent said they had anticipated other benefits. Of these, most said that the wind farm would benefit the local economy, which translates into 10% of respondents overall saying they thought the wind farm would be beneficial to the local community. This proportion increases to 17% of those who live in the high proximity zone.

2.5.5 Respondents were then asked which of the potential benefits they had actually experienced and their responses are presented in Table 9. The table shows that respondents actually experienced some of the benefits they anticipated. Overall, forty-nine per cent of respondents said they had not experienced any benefit from the local wind farm. This decreased to 43% in the high proximity zone and increased to 52% in the low proximity zone. Eighteen per cent overall say that local jobs have been created and this increases to 26% of those in the high proximity zone.

Table 9: Actual benefits experienced by respondents (%)

Actual benefits	High proximity	Medium proximity	Low proximity	Overall
Cheaper electricity	9	6	1	**6**
Local jobs created	26	15	5	**18**
None	43	57	52	**49**
Electricity supplied locally	8	8	2	**7**
Reduced pollution	17	12	3	**12**
Increased tourism	11	6	1	**7**
Other	9	1	1	**5**
Don't know	11	16	37	**19**
Base	215	108	107	**430**

Note for table
Nil per cent is indicated by a dash (-); figures between 0% and 0.5% are indicated by an asterisk (*); figures between 0.5% and 1% are rounded up to 1%. Figures may add up to more than 100% as respondents could give more than one answer.

2.5.6 There were however, differences between the anticipated and the actual benefits. These differences are shown in Figure 3 for all respondents. In relation to each of the potential benefits, the actual proportion who experienced the benefit is lower than the proportion who anticipated the benefit. The biggest difference between anticipated and actual benefits is in relation to cheaper electricity and the smallest difference is in relation to increased tourism.

Figure 3: Anticipated and actual benefits overall (%) Base = 430

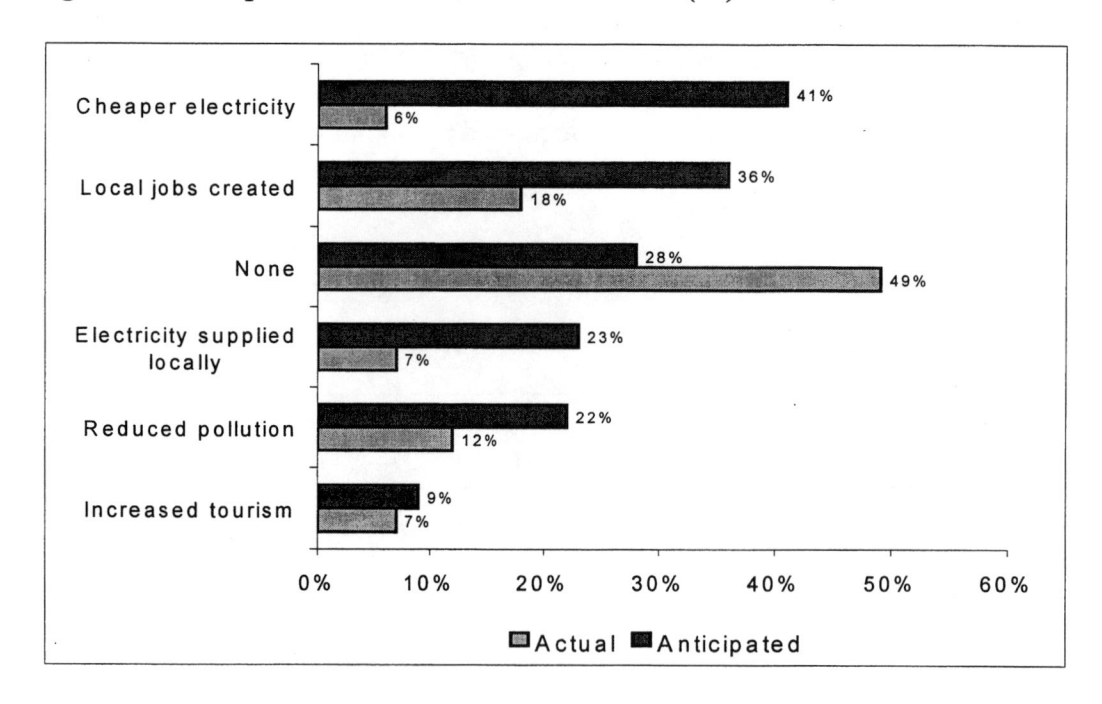

Benefit	Actual	Anticipated
Cheaper electricity	6%	41%
Local jobs created	18%	36%
None	49%	28%
Electricity supplied locally	7%	23%
Reduced pollution	12%	22%
Increased tourism	7%	9%

2.5.7 The gap between anticipated and actual benefits, especially reduced pollution and the creation of local jobs, is smaller in the high proximity zone than in the other two zones as illustrated in Figure 4. Actual experience of increased tourism is higher than anticipated in the high proximity zone.

Figure 4: Anticipated and actual benefits in the high proximity zone (%) Base = 215

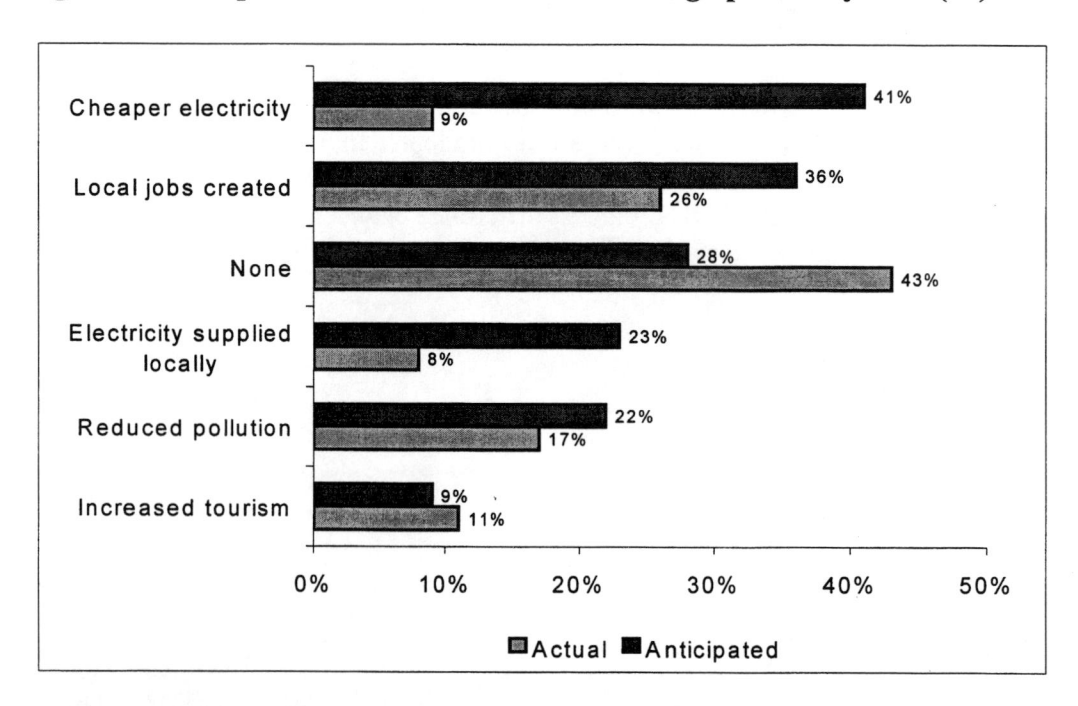

Benefit	Actual	Anticipated
Cheaper electricity	9%	41%
Local jobs created	26%	36%
None	43%	28%
Electricity supplied locally	8%	23%
Reduced pollution	17%	22%
Increased tourism	11%	9%

2.5.8 In the case of the 'other benefits', discussed above, five percent thought that the wind farm had been of benefit to the local community. In three out of the four wind farm areas, the wind farm developers pay an annual sum of money to the local community councils in the local area to be used for the benefit of the local area. It is probable that this is a factor influencing residents' views of whether the farm has been of benefit to the local community. Several respondents mentioned examples such as the village hall getting a new roof and improvements to the local church.

2.6 SOURCES OF INFORMATION AND COMMUNICATION

2.6.1 The data and discussion above show that there are significant differences between expectations and the reality of having a wind farm in the local area. It is probable that anticipated problems and benefits are influenced by residents' sources of information about wind farms. Figure 5 shows how respondents were informed about the wind farm when it was first proposed. Overall, a large proportion (36%) don't know or can't remember where they learnt about the wind farm. This decreases to 30% in the high proximity area and increases to 51% in the low proximity zone.

2.6.2 The most common source of information was from local newspapers (42%), followed by 'word of mouth' or local people (8%) and then the local authority (7%) and the developer (7%). The proportions who gained information from the local authority or the developer increased slightly in the high proximity zone to 10% and 9% respectively.

Figure 5: Sources of information about the wind farm when it was proposed (%)

Base = 377 Those living in the area when the farm was proposed

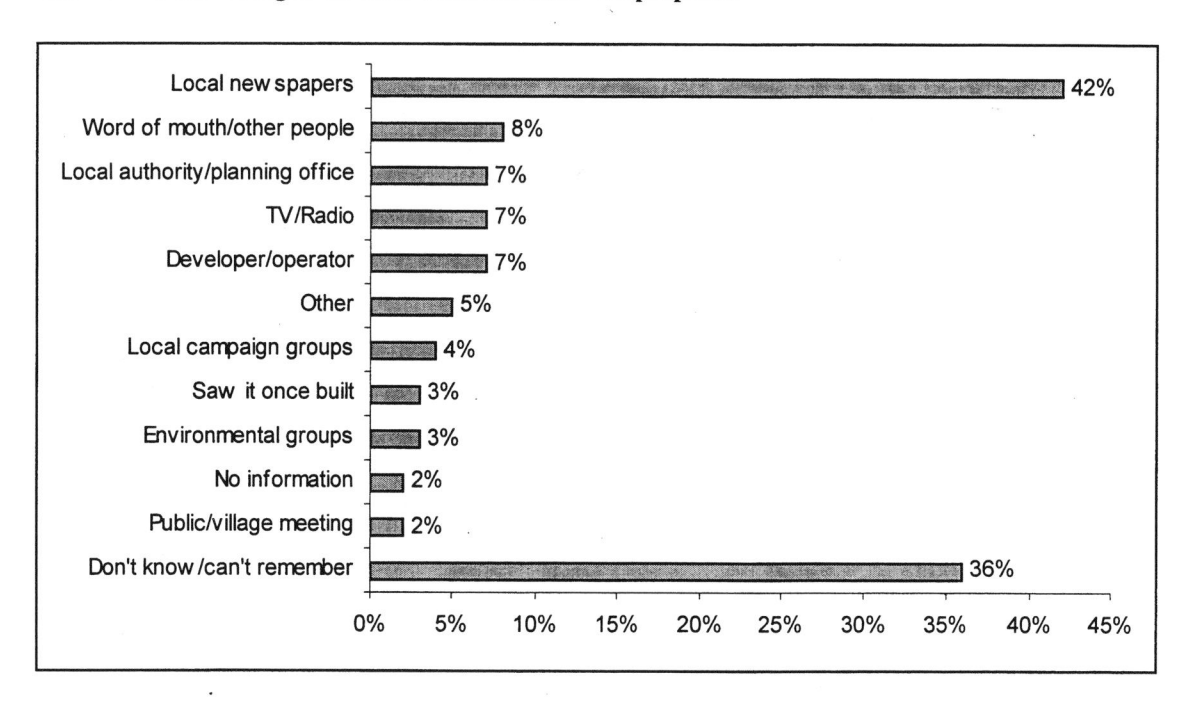

Source	%
Local newspapers	42%
Word of mouth/other people	8%
Local authority/planning office	7%
TV/Radio	7%
Developer/operator	7%
Other	5%
Local campaign groups	4%
Saw it once built	3%
Environmental groups	3%
No information	2%
Public/village meeting	2%
Don't know /can't remember	36%

2.6.3 Respondents were asked whether the wind farm developer or the local authority planning office conducted any public consultation. Tables 10 and 11 show that there is a fairly high lack of awareness as to whether consultation did take place and a large proportion of respondents who say there was no consultation.

Table 10: Awareness of consultation by developer (%)

Consultation by developer	High proximity	Medium proximity	Low proximity	Overall
Yes	24	25	5	**20**
No	41	45	54	**45**
Don't know/not stated	34	30	41	**35**
Base	198	96	83	377

Note for table
Nil per cent is indicated by a dash (-); figures between 0% and 0.5% are indicated by an asterisk (*); figures between 0.5% and 1% are rounded up to 1%.

2.6.4 Overall, 20% of respondents said that the developer had conducted consultation with the community. Residents in the high and medium proximity zones were more likely to say that the developer consulted with the public. Of the 20% who said the developer had conducted public consultation (n=76), 64% were satisfied with the level of consultation.

Table 11: Awareness of consultation by local authority (%)

Consultation by local authority	High proximity	Medium proximity	Low proximity	Overall
Yes	19	22	5	**17**
No	70	65	77	**70**
Don't know/not stated	11	14	18	**13**
Base	198	96	83	377

Note for table
Nil per cent is indicated by a dash (-); figures between 0% and 0.5% are indicated by an asterisk (*); figures between 0.5% and 1% are rounded up to 1%.

2.6.5 Only 17% of respondents said that the local authority carried out public consultation. The proportions aware of consultation were higher in the high and medium proximity zones that in the low proximity zone, as might be expected. Of those respondents who were aware of Local Authority consultation (n=63), 59% were satisfied with the level of consultation.

2.6.6 The survey data does not suggest that residents' perceptions of benefits and problems were affected by the level of consultation or by sources of information. This is, however, largely due to the fact that the numbers are too small in some of the categories to measure any differences.

2.7 ATTITUDES TOWARDS POSSIBLE FUTURE DEVELOPMENT

2.7.1 A good indicator of residents' feelings towards their local wind farm is their attitude towards proposed further development. Respondents were asked how concerned they would be if there was a proposal to add more turbines to the existing wind farm.

Table 12: Level of concern if more turbines were added to existing wind farm (%)

Level of concern	High proximity	Medium proximity	Low proximity	Overall
Very concerned	4	6	4	**4**
Fairly concerned	12	7	7	**10**
Not very concerned	26	17	21	**22**
Not at all concerned	57	69	60	**60**
Don't know/ not stated	2	1	7	**3**
Base	215	108	107	**430**

Note for table
Nil per cent is indicated by a dash (-); figures between 0% and 0.5% are indicated by an asterisk (*); figures between 0.5% and 1% are rounded up to 1%.

2.7.2 Table 12 shows that, overall, 82% said they would either be not very or not at all concerned about extra turbines. A slightly higher proportion of residents in the high-proximity zone (16%) would be concerned about further developments than in the other two zones.

2.7.3 Tables 13 and 14 examine whether the circumstances and frequency whereby residents currently see the wind farm affect their level of concern about future wind farm development.

Table 13: Level of concern if extra turbines were to be added by frequency of seeing the wind farm (%)

Level of concern	Every Day	Most Days	Occasionally	Never	Total
Very concerned	6	6	4	4	**4**
Fairly concerned	9	12	10	7	**10**
Not very concerned	9	22	25	26	**22**
Not at all concerned	76	58	58	54	**61**
Don't know/not stated	-	2	3	9	**3**
Base	66	67	240	57	**430**

Note for table
Nil per cent is indicated by a dash (-); figures between 0% and 0.5% are indicated by an asterisk (*); figures between 0.5% and 1% are rounded up to 1%.

2.7.4 Table 13 shows that a high proportion of residents who see the wind farm 'every day' or 'most days' would be largely unconcerned if more turbines were added to the wind farm. Those residents who see the wind farm most frequently are most likely to be 'not at all concerned' (76%).

Table 14: Level of concern if extra turbines were proposed by circumstances in which respondent sees the wind farm (% of responses)

Level of concern	From home or garden	Travelling on local roads	Travelling on major roads	Out hill walking/ in the country	Overall
Very concerned	10	7	4	5	**4**
Fairly concerned	10	10	12	6	**10**
Not very concerned	10	21	22	17	**22**
Not at all concerned	69	62	60	73	**61**
Don't know/not stated	-	-	2	0	**3**
Base	48	170	245	66	**430**

Note for table
Nil per cent is indicated by a dash (-); figures between 0% and 0.5% are indicated by an asterisk (*); figures between 0.5% and 1% are rounded up to 1%.

2.7.5 The circumstances whereby the wind farm is seen by respondents has slightly more influence on their level of concern. Twenty percent of residents who can see the wind farm from their home or garden would be concerned if more turbines were to be added compared with 14% overall.

2.7.6 Adding more turbines to an existing wind farm might perhaps not be of concern to residents as the additional effect might not be that great. A proposal to build another wind farm in another part of the local area could potentially cause more concern. Table 15 shows that almost twice as many respondents (27%) would be concerned about proposals for another wind farm than would be concerned about extra turbines to the existing farm (14%).

Table 15: Level of concern if another wind farm was proposed for the local area (%)

Level of concern	High proximity	Medium proximity	Low proximity	Overall
Very concerned	5	10	7	**7**
Fairly concerned	23	15	19	**20**
Not very concerned	29	26	36	**30**
Not at all concerned	39	47	31	**39**
Don't know/ not stated	5	2	7	**5**
Base	215	108	107	**430**

Note for table
Nil per cent is indicated by a dash (-); figures between 0% and 0.5% are indicated by an asterisk (*); figures between 0.5% and 1% are rounded up to 1%.

2.7.7 However, overall, 69% of respondents would be either not very or not at all concerned about another wind farm in the area. A slightly higher proportion of those in the medium proximity zone would not be concerned compared with the other two zones. Table 16 shows that the frequency with which respondents see the wind farm does not have any major effect on levels of concern.

Table 16: Level of concern if another wind farm was proposed by frequency of seeing the wind farm (%)

Level of concern	Every Day	Most Days	Occasionally	Never	Overall
Very concerned	14	8	6	4	**7**
Fairly concerned	14	22	21	18	**20**
Not very concerned	20	31	33	25	**30**
Not at all concerned	52	33	36	42	**39**
Don't know/not stated	2	6	3	12	**5**
Base	66	67	240	57	**430**

Note for table
Nil per cent is indicated by a dash (-); figures between 0% and 0.5% are indicated by an asterisk (*); figures between 0.5% and 1% are rounded up to 1%.

2.7.8 Again, the circumstances whereby respondents see the wind farm appears to have slightly more influence on their level of concern than the frequency of sightings. Residents who can see the wind farm from their home or garden are more likely than those who see it in other circumstances to be concerned about a proposal for another wind farm – 36% compared with 27% overall (Table 17).

Table 17: Level of concern if another wind farm was proposed by circumstances in which respondent sees the wind farm (%)

Level of concern	From home or garden	Travelling on local roads	Travelling on major roads	Out hill walking/ in the country	Overall
Very concerned	13	10	9	12	7
Fairly concerned	23	18	20	14	**20**
Not very concerned	21	32	32	26	**30**
Not at all concerned	44	39	35	46	**39**
Don't know/not stated	-	2	4	3	**5**
Base	48	170	245	66	**430**

Note for table
Nil per cent is indicated by a dash (-); figures between 0% and 0.5% are indicated by an asterisk (*) figures between 0.5% and 1% are rounded up to 1%.

2.7.9 Respondents were asked what kind of locations they would prefer for wind farm developments. Figure 6 shows that 48% said wind farms should not be situated in inhabited places and 36% that they should be high on hills. There was very little variation across the different proximity zones although respondents in the high proximity zone were slightly more likely to say that wind farms should be located high on hills. Those who see the wind farm daily are more likely than those who see the wind farm less frequently, to say that they should not be located in inhabited areas.

Figure 6: Preferred locations for wind farms (%) Base = 430

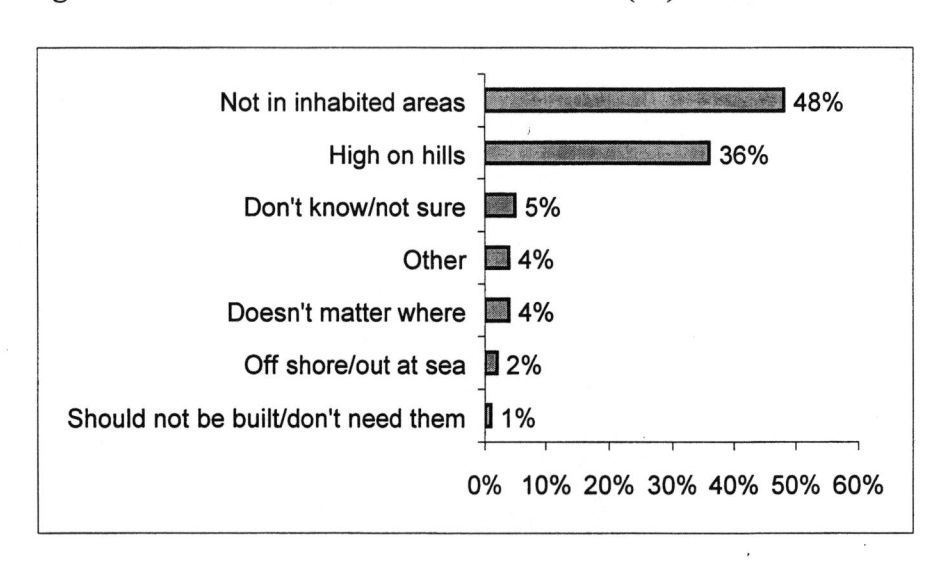

- Not in inhabited areas: 48%
- High on hills: 36%
- Don't know/not sure: 5%
- Other: 4%
- Doesn't matter where: 4%
- Off shore/out at sea: 2%
- Should not be built/don't need them: 1%

2.8 LEVELS OF ENVIRONMENTAL CONCERN

2.8.1 A factor which may effect peoples' attitudes towards the potential merits or problems with wind farms is their level of interest or concern about environmental conservation. It is possible that a high level of concern for protecting the environment might increase *support* for wind farms due to concerns about the effect of nuclear or fossil fuel. On the other hand, concern for the environment might also increase *opposition* to wind farms, due to a concern about protecting the landscape.

2.8.2 An indicator of environmental concern was calculated for each respondent based on their responses to certain questions in the survey (see questions 24 to 29 in Appendix Two). One point was awarded for a positive response for each of these questions enabling a score on a scale of between zero and six to be calculated. A score of zero indicates a very low level of concern about the environment and six a very high level of concern.

2.8.3 Overall, 9% of respondents are members of organisations associated with the environment or conservation and 74% use recycling facilities. Eighty six per cent were either fairly or very concerned about 'damage to the countryside' and 'loss of wildlife'. Seventy two per cent were concerned about 'global warming and climate change' or 'depletion of natural resources such as coal, oil or gas'.

2.8.4 The relationship between environmental concern and support for wind farms is not simple. Low levels of environmental concern can be associated with indifference towards wind farm developments or problems with wind farms. Conversely, high levels of environmental concern can be associated with opposition to wind farms.

Table 18: Level of concern if another wind farm was proposed by environmental concern score (%)

	0-2	3-4	5-6	Overall
Very concerned	-	5	10	7
Fairly concerned	9	20	23	20
Not very concerned	26	32	29	30
Not at all concerned	62	35	34	39
Don't know/not stated	3	7	4	5
Base	65	164	201	430

Note for table
Nil per cent is indicated by a dash (-); figures between 0% and 0.5% are indicated by an asterisk (*); figures between 0.5% and 1% are rounded up to 1%.

2.8.5 Table 18 shows an example of this complex relationship, whereby respondents who score highest on the environmental indicator scale are more likely to be concerned about another wind farm development in the area than those who score lowest on the environmental indicator. This pattern emerges at other questions and illustrates that the relationship between environmental concern and attitudes towards wind farms is complex.

CHAPTER THREE: CONCLUSIONS

3.1 Respondents appear to be generally positive about the local wind farm. When asked what, if anything they liked about the wind farm, only 11% said that there was nothing they liked, with a further 21% saying that they did not know if there was anything they did not like about the farm. Overall, 67% of respondents liked something about the wind farm with this proportion increasing to 73% of those living in the high proximity zone. Generally, those living closest to the wind farm were more likely to mention positive aspects of the wind farm when asked this question. Similarly, those who see the wind farm most often were more likely to give positive responses when asked what they liked about the wind farm.

3.2 The visual impact of wind farms does not appear to be a major issue for local residents. Although all respondents live within 20km of a wind farm, only 13% can see it from their home or garden and only 11% see the farm every day. Only 10% overall said that they did not like the wind farm because it was unsightly or spoilt the view. Similarly, although 27% of respondents had expected the landscape to be spoilt by the wind farm, only 5% said they had actually experienced this after the wind farm was developed.

3.3 Noise did not feature as an issue for residents either. Overall, 2% said that they disliked the wind farm because it was noisy. Although 12% of respondents had expected to experience a problem with noise, only 1% had actually experienced noise as a problem.

3.4 None of the respondents mentioned the problem of 'shadow flicker' or 'driver distraction' when asked what they disliked about wind farms. Only 2% of all respondents had experienced problems with TV or radio interference.

3.5 Generally, the proportion of respondents who anticipated problems was far higher than the proportion who actually experienced problems. Overall, 40% expected problems and 9% actually experienced problems. This suggests that respondents expectations were based on perception rather than on experience.

3.6 However, the proportion of respondents who expected benefits was higher than the proportion who actually experienced them. Overall, 63% expected at least one benefit while 32% say they had experienced a benefit.

3.7 Respondents showed low levels of awareness of consultation by either the wind farm developer or the local authority. Twenty percent were aware of consultation being conducted by the wind farm developer and 17% were aware of consultation by the local authority. Additionally, the most common source of information about the wind farm was 'local newspapers', with only 7% receiving information from the local authority or the wind farm developer.

3.8 As the number of people aware of consultation was so low, the survey data could not adequately examine whether respondents' attitudes were affected by consultation or sources of information about the wind farm. However, as there appears to be a significant gap between perception and experience, increased consultation and information provision about wind farms might reduce the amount of problems anticipated by local residents.

3.9 The generally positive attitude of respondents is reflected in the fact that overall, only 14% of respondents would be concerned if extra turbines were added to the existing wind farm. However, a higher proportion of respondents who see the farm from their home or garden would be concerned if more turbines were added (20%).

3.10 A higher proportion of respondents would be concerned if another wind farm were proposed for the local area (27%). This proportion is similar for the three proximity zones and by frequency of sighting. A higher proportion of those who can see the wind farm from their home or garden would be concerned about another wind farm in the local area (33%).

3.11 These findings suggest that, although a high proportion of residents would not be concerned about additional turbines or additional wind farms, it would perhaps be preferable to site them where they can not be seen from peoples' homes.

3.12 Respondents demonstrate positive attitudes towards their local wind farms but when asked where wind farms should be located, the majority said they should be in uninhabited areas high on hills.

3.13 The majority of respondents currently living near wind farms have not experienced any problems with the wind farms. The problems they had anticipated did not materialise in the vast majority of cases (only 9% experienced any problems compared with 40% who expected them). This suggests that the information provided about wind farms and the explanations for their development are crucial in order to reduce anticipation of problems.

APPENDIX ONE: BACKGROUND INFORMATION ABOUT THE WIND FARMS

Background information on the individual wind farms and the associated planning and development process was gathered from the relevant local authority planning departments. In particular, information was gathered on the size and visual impact of the wind farms, the level of public support or objection during the planning process, and any planning conditions. This provides a context for examining the residents attitudes towards the wind farms by, for example giving an idea of their relative size.

Hagshaw Hill, South Lanarkshire

This is the oldest wind farm in Scotland, with planning permission applied for in May 1994 and approved in February 1995. The farm has been operational since November 1995 and consists of 26 Bonus 600 turbines.

Overall, the planning application went smoothly, there were only two letters of representation from the public. One from a local resident who was concerned about noise and depopulation of the area and one from an environmental organisation concerned about the effects on bird life.

During consultation, the application was found to conform to national planning guidelines (NPPG6 published August 1994) and fit in with the local and structure plans. The visual and ecological effects were found acceptable. No further objections were raised and no further planning conditions were required except for a limit on the number of turbines (30).

The wind farm is fairly visible and can be see for up to 10 miles but only from certain places. Many local residents will not see the wind farm regularly due to the siting of the farm and the local landform.

Windy Standard, New Galloway

Planning permission was granted in November 1995 and the farm has been operational since September 1996 and consists of 36 Nordtank 600 turbines.

The planning authority said that the planning application was aided by the willingness of the developer to undertake consultation and listen to representations. The developer conducted a lot of pre-application consultation and environmental impact studies including a bird survey before and after development.

The site for Windy Standard aided the application as it has very low visibility and is only highly visible from the hills. A Section 50 (now 75) agreement was arranged whereby a fund has been set aside for the reinstatement of the land should this be necessary.

There were however many more letters of representation in this case than in the case of Hagshaw Hill with a total of 52 letters received. Of these letters, 4 were from local residents,

3 from residents elsewhere in the area and 45 were from people living England and Wales. There were also four letters of support for the wind farm project.

After development, local residents were invited to an open day which was very well attended. There have been no representations since the farm was operational about any noise or other problems.

In an arrangement separate from the planning application, the developer contributes funds to local community councils during the life of the wind farm. This was not a material consideration of the planning process.

Novar, Highlands

Planning consent was achieved in December 1995 and Novar has been operational since October 1997 and consists of 34 Bonus 500 turbines and the planning authority describes the development as a success. The process was helped by the behaviour of the developer which was very good. They prepared an environmental statement, were willing to take comments on board and were very co-operative both during and after the process. There was an open day at the farm soon after opening which was attended by a large number of visitors.

A small number of letters of letters of representation were received from local people. Overall 8 objections were received during the consultation period – 6 were from local residents, 1 from an environmental organisation and 1 from a representative of an archaeological site. A further 6 objections were received too late – 2 of these were from and environmental campaign group opposed to wind farm development and 4 from local residents.

According to the planning authority, a concern among local residents was in relation to TV reception. The developer conducted a study to assess the impact and consulted with the BBC who said they did not envisage any major problems. The developer made a commitment to sort out any problems that residents did experience with reception. A few residents did experience problems with television and radio reception and this was remedied by the developer.

Another issue arising among a few of the local residents was the feeling that the wind farm was un-necessary. Electricity supplies were already perceived to be local and renewable as the supplies are largely hydro-electric. There was a feeling that the wind farm was to be of benefit to people outside the area rather than local people. This did not feature heavily in the survey results although a few respondents did say that the farms were not necessary.

This wind farm is very visible but only from certain places, people are most likely to see it whilst driving on the A9 and then only for a while. It is visible from maybe 30km away but only if you know where to look and what you are looking for. There have been no complaints to the local authority about the wind farm.

In an arrangement separate from the planning application, the developer contributes funds to local community councils during the life of the wind farm. This was not a material consideration of the planning process.

Beinn Glas, Argyll and Bute

Beinn Glas is the newest and smallest of the four wind farms. Planning consent was given in November 1997 and it has been operational since May 1999 and consists of only fourteen turbines. Sixty letters of objection and forty letters of support were received during the planning consultation.

Two turbines out of the original sixteen were deleted from the proposals. This amendment resulted from negotiations between the Planning Authority and the developer and was not as a direct result of the public representations. The reason for the amendment was landscape impact. There were a large number of planning conditions and a Section 75 agreement.

The Beinn Glas developer had an open day to which local residents were invited and this was well attended.

In the opinion of the Planning Authority the visual impact is an acceptable compromise between landscape conservation and development of renewable power technologies.

In an arrangement separate from the planning application, the developer contributes funds to local community councils during the life of the wind farm. This was not a material consideration of the planning process.

APPENDIX TWO: THE QUESTIONNAIRE

INTRODUCTION

Good morning/afternoon, I'm calling from System Three on behalf of the Scottish Executive. The Scottish Executive is interested in what people who live in Scotland think about wind farms. {Interviewer – select person over 16 with the next birthday}.

Would you mind answering some questions? All your answers will be strictly confidential. No one will see your individual responses and no information will be generated that would enable any individual to be identified.

Q 1 First, can I just check whether this is your sole or main residence or whether it is a second or holiday home?

	(18)	
Permanent residence	1	**A3**
Holiday/second home	2	A2

Q 2 Do you or a member of your family own the property?

	(19)	
Yes	1	A-3
No	2	DO NOT INTERVIEW

Q 3 And how long have you been in this property?

Write in with preceding zeros (20)(21) ☐☐ Years (22)(23) ☐☐ Months

Q 4 What, if anything do you like about having a wind farm in your local area? PROBE FULLY

_____ (24)

_____ (25)

_____ (26)

_____ (27)

Q 5 What if anything do you dislike about having a wind farm in your local area?

_____ (28)

_____ (29)

_____ (30)

_____ (31)

Q 6 And how often do you see the wind farm?

		(32)	
Every day	1		
Most days	2	Q7
Occasionally	3	
Never	3	Q8

Q 7 In what circumstances would you see the wind farm. Can you see it READ OUT

	(33)
From your home	1
When travelling on local road (in the town or village)	2
When travelling on major roads	3
Out walking in the country	4
In other circumstances (write in) _____	5

Q 8 When you first learnt about the wind farm, did you think there might be a problem with any of the following? READ OUT

	Yes	NO	N/A	
Noise from the turbines	1	2		(34)
The look of the landscape being spoiled	1	2		(35)
Interference with TV and radio reception	1	2		(36)
Damage to plants or animals	1	2		(37)
Noise or disturbance during construction	1	2	3	(38)
Extra traffic during construction	1	2	3	(39)
A reduction in house prices	1	2		(40)

Q 9   Was there anything else you thought might be a problem when you first learnt about the wind farm? PROBE, Anything else?

_____ (41)

_____ (42)

_____ (43)

Q 10   Which, if any of these things would you say is a problem you have actually experienced with the wind farm? CODE ALL THAT APPLY

	(44)
Noise from the turbines	1
The look of the landscape being spoiled	2
Interference with TV and radio reception	3
Damage to plants or animals	4
Noise or disturbance during construction	5
Extra traffic during construction	6
A reduction in house prices	7
	(45)
Others (specify) _____	1
None	0
DK	Y

Q 11   Still thinking about when you first learnt about the wind farm, which, if any of the following did you expect? READ OUT. CODE ALL THAT APPLY

	(46)
Reduced pollution	1
Cheaper electricity	2
Electricity supplied locally	3
Local jobs created	4
Increased tourism	5
None	0
DK	Y

Q 12   Was there anything else that you thought might be good or beneficial about having a wind farm in the local area?

_____ (47)

_____ (48)

_____ (49)

Q 13 Which, if any of those things would you say has turned out to be a benefit of having a local wind farm? CODE ALL THAT APPLY

	(50)
Reduced pollution	1
Cheaper electricity	2
Electricity supplied locally	3
Local jobs created	4
Tourists or visitors	5
Others (specify) _____	6
None of these	0
DK	Y

Q 14 Was the wind farm already here when you moved in or has it been built since then?

	(51)	
Farm already here	1	QA-21
Built since then	2	QA-15

Q 15 When the wind farm was first proposed, which sources did you get information from about the wind farm? CODE ALL THAT APPLY

	(52)
The developer/operator	1
Local newspapers	2
TV or radio	3
Environmental groups	4
Local campaign groups	5
The local authority/ planning office	6
Other (write in) _____	7
Can't remember/DK	Y

Q 16 Did the developer conduct any public consultation about the wind farm?

	(53)	
Yes	1	QA-17
No	2	QA-18
DK	Y	

Q 17 How satisfied were you with the level of consultation by the developer? Would you say that you were READ OUT

	(54)
Very satisfied	1
Fairly satisfied	2
Neither satisfied nor dissatisfied	3
Fairly dissatisfied	4
Very dissatisfied	5

Q-18 Did the local authority planning department conduct any public consultation?

	(55)	
Yes	1	QA-19
No	2	GO TO INT AFTER Q-19

Q-19 And how satisfied were you with the level of consultation by the local authority? Would you say you were, READ OUT

	(56)	
Very satisfied	1	
Fairly satisfied	2	QA-21
Neither satisfied nor dissatisfied	3	
Fairly dissatisfied	4	QA-20
Very dissatisfied	5	

> IF CODED 4 OR 5 AT QA-17 OR QA-19, CONTINUE. OTHERS GO TO QA-21

Q-20 Why were you dissatisfied with the level of consultation?

_____ (57)

_____ (58)

_____ (59)

ASK ALL

Q-21 How would you feel if more turbines were to be added to the existing wind farm? Would you be … READ OUT

[Interviewer – if asked, say 2 or 3 more turbines]

	(60)
Very concerned	1
Fairly concerned	2
Not very concerned	3
Not at all concerned	4
DK	Y

Q -22 And how would you feel if another wind farm was proposed in your local area? Would you be …. READ OUT

[Interviewer – if asked, say within 2 or 3 miles]

	(61)
Very concerned	1
Fairly concerned	2
Not very concerned	3
Not at all concerned	4
DK	Y

Q 23 In what kinds of places do you think wind farms should be located?

_____ (62)

_____ (63)

Q -24 Do you use recycling facilities such as bottle banks and paper banks?

	(64)
Yes	1
No	2

Q -25 Are you a member of any organisations that are associated with conservation or the environment? CODE ALL THAT APPLY.

	(65)
Greenpeace	1
Friends of the Earth	2
National Trust for Scotland	3
Scottish Wildlife Trust	4
World Fund for Nature (World Wildlife Fund)	5
British Trust for Nature Conservation	6
RSPB	7
Local amenity group	8
Other (write in) _____	9
None	0

Q 26 How concerned are you about damage to the countryside?

	(66)
Very concerned	1
Fairly concerned	2
Not very concerned	4
Not at all-concerned	5
DK	Y

Q -27 How concerned are you about loss of wildlife?

	(67)
Very concerned	1
Fairly concerned	2
Not very concerned	4
Not at all-concerned	5
DK	Y

Q -28 How concerned are you about global warming or climate change?

	(68)
Very concerned	1
Fairly concerned	2
Not very concerned	4
Not at all-concerned	5
DK	Y

Q 29 How concerned are you about the depletion of natural resources such as coal, oil or gas?

	(69)
Very concerned	1
Fairly concerned	2
Not very concerned	4
Not at all-concerned	5
DK	Y

Q -30 Can you tell what age you are please?

(70) (71)

Write in with preceding zeros ☐☐ Years

Q -31 Can you tell me how many people aged 16 or over live in your household? (WRITE IN)

(72)
☐

Q 32 And can you tell me how many children (aged 15 or under) living in your households? **(CODE NUMBER BELOW – IF NONE, CODE '0')**

(73)
☐

CRU RESEARCH - PUBLICATIONS LIST FROM 1999

Poor Housing and Ill Health: A Summary of Research Evidence: Housing Research Branch. (1999) (£2.50)

One Stop Shop Arrangements for Development Related Local Authority Functions: Centre for Planning Research, School of Town and Regional Planning, University of Dundee. (1999) (£5.00)
Summary available: Development Department Research Findings No.63

Research on Walking: System Three. (1999) (£5.00)

Resolving Neighbour Disputes Through Mediation in Scotland: Centre for Criminological and Legal research, University of Sheffield. (1999) (£4.00)
Summary available: Development Department Research Findings No.64

Literature Review of Social Exclusion: Centre for Urban and Regional Studies, University of Birmingham. (1999) (£5.00)

Mentally Disordered Offenders and Criminal Proceedings: Dr M Burman, Department of Sociology and Ms C Connelly, School of Law, University of Glasgow. (1999) (£7.50)

Evaluation of Experimental Bail Supervision Schemes: Ewen McCaig and Jeremy Hardin, MVA Consultancy. (1999) (£6.00)
Summary available: Social Work Research Findings No.28

An Evaluation of the 1997/98 Keep Warm This Winter Campaign: Simon Anderson and Becki Sawyer, System 3. (1999) (£5.00)
Summary available: Social Work Research Findings No.29

Attitudes Towards Crime, Victimisation and the Police in Scotland: A Comparison of White and Ethnic Minority Views: Jason Ditton, Jon Bannister, Stephen Farrall & Elizabeth Gilchrist`, Scottish Centre for Criminology. (1999) (£5.00)
Summary available: Crime and Criminal Justice Research Findings No.28

The Safer Cities Programme in Scotland – Evaluation of the Aberdeen (North East) Safer Cities Project: MVA. (1999) (£5.00)

Review of National Planning Policy Guidelines: Land Use Consultants. (1999) (£5.00)
Summary available: Development Department Research Findings No.65

Development Department Research Programme 1999-2000. (1999) (Free)

Environment Group Research Programme 1999-2000. (1999) (Free)

Rural Policy Research Programme 1999-2000. (1999) (Free)

Referrals between Advice Agencies and Solicitors: Carole Millar Research. (1999) (£5.00)
Summary available: Legal Studies Research Findings No.21

Life Sentence Prisoners in Scotland: Diane Machin, Nicola Coghill, Liz Levy. (1999) (£3.50)
Summary available: Crime and Criminal Justice Research Findings No.29

Report on a Conference on Domestic Violence in Scotland, Scottish Police College, Tulliallan: The Scottish Office, The Health Education Board for Scotland, The Convention of Scottish Local Authorities, The Scottish Needs Assessment Programme. (1999) (£5.00)

Making it Safe to Speak? Witness Intimidation and Protection in Strathclyde: Nicholas Fyfe, Heather McKay, University of Strathclyde. (1999) (£7.50)

Supporting Court Users: The Pilot In-Court Advice Project in Edinburgh Sheriff Court: Elaine Samuel, Department of Social Policy, University of Edinburgh. (1999) (£5.00)
Summary available: Legal Studies Research Findings No. 22

The Role of Mediation in Family Disputes in Scotland: Jane Lewis, Social and Community Planning Research. (1999) (£5.00)
Summary available: Legal Studies Research Findings No. 23

Research on Women's Issues in Scotland: An Overview: Esther Breitenbach. (1999) (Free)
Summary only available: Women's Issues Research Findings No. 1

Women in Decision-Making in Scotland: A Review of Research: Fiona Myers, University of Edinburgh. (1999) (Free)
Summary only available: Women's Issues Research Findings No. 2

Evaluation of the Debtors (Scotland) Act 1987: Study of Individual Creditors: Debbie Headrick and Alison Platts. (1999) (£5.00)
Summary available: Legal Studies Research Findings No. 10

Evaluation of the Debtors (Scotland) Act 1987: Study of Commercial Creditors: Alison Platts. (1999) (£5.00)
Summary available: Legal Studies Research Findings No. 11

Evaluation of the Debtors (Scotland) Act 1987: Study of Debtors: David Whyte. (1999) (£5.00)
Summary available: Legal Studies Research Findings No. 12

Evaluation of the Debtors (Scotland) Act 1987: Study of Facilitators: Andrew Fleming. (1999) (£5.00)
Summary available: Legal Studies Research Findings No. 13

Evaluation of the Debtors (Scotland) Act 1987: Survey of Poindings and Warrant Sales: Andrew Fleming. (1999) (£5.00)
Summary available: Legal Studies Research Findings No. 14)

Evaluation of the Debtors (Scotland) Act 1987: Survey of Payment Actions in the Sheriff Court: Andrew Fleming, Alison Platts. (1999) (£5.00)
Summary available: Legal Studies Research Findings No. 15

Evaluation of the Debtors (Scotland) Act 1987: Analysis of Diligence Statistics: Andrew Fleming, Alison Platts. (1999) (£5.00)
Summary available: Legal Studies Research Findings No. 16

Evaluation of the Debtors (Scotland) Act 1987: Overview: Alison Platts. (1999) (£5.00)

Looking After Children in Scotland: Susanne Wheelaghan, Malcolm Hill, Moira Borland, Lydia Lambert and John Triseliotis. (1999) (£5.00)
Summary available: Social Work Research Findings No.30

The Evaluation of Children's Hearings in Scotland: Children in Focus: Lorraine Waterhouse, Janice McGhee, Nancy Loucks, Bill Whyte & Helen Kay
Summary available: Social Work Research Findings No.31

Taking Account of Victims in the Criminal Justice System: A Review of the Literature: Andrew Sanders. (1999) (£5.00)
Summary available: Social Work Research Findings No.32

Social Inclusion Bulletin No.1: (1999) (Free)

Geese and their Interactions with Agriculture and the Environment: JS Kirby, M Owen & JM Rowcliffe. (1999) (£10.00)
Summary available: Countryside and Natural Heritage Research Findings No.1

The Recording of Wildlife Crime in Scotland: Ed Conway. (1999) (£10.00)
Summary available: Countryside and Natural Heritage Research Findings No.2

Socio-Economic Benefits from Natura 2000: GF Broom, JR Crabtree, D Roberts & G Hill. (1999) (£5.00)
Summary available: Countryside and Natural Heritage Research Findings No.3

Crime and the Farming Community: The Scottish Farm Crime Survey 1998: Andra Laird, Sue Granville & Ruth Montgomery. (1999) (£10.00)
Summary available: Agricultural Policy Co-ordination and Rural Development Research Findings No.1

New Ideas in Rural Development No 7: Community Development Agents in Rural Scotland: Lynn Watkins & Alison Brown. (1999) (£2.50)
Summary available: Agricultural Policy Co-ordination and Rural Development Research Findings No.2

New Ideas in Rural Development No 8: Tackling Crime in Rural Scotland: Mary-Ann Smyth. (1999) (£2.50)
Summary available: Agricultural Policy Co-ordination and Rural Development Research Findings No.3

Study of the Impact of Migration in Rural Scotland: Professor Allan Findlay, Dr David Short, Dr Aileen Stockdale, Anne Findlay, Lin N Li, Lorna Philip. (1999) (£10.00)
Summary available: Agricultural Policy Co-ordination and Rural Development Research Findings No.4

An Electoral System for Scottish Local Government: Modelling Some Alternatives: John Curtice. (1999) (£5.00)

Writing for the CRU Research Series: Ann Millar, Sue Morris & Alison Platts. (1999) (Free)

The Effect of Closed Circuit Television on Recorded Crime Rates and Public Concern about Crime in Glasgow: Jason Ditton, Emma Short, Samuel Phillips, Clive Norris & Gary Armstrong. (1999) (£5.00)
Summary available: Crime and Criminal Justice Research Findings No.30

Working with Persistent Juvenile Offenders: An Evaluation of the Apex Cueten Project: David Lobley & David Smith. (1999) (£5.00)
Summary available: Crime and Criminal Justice Research Findings No.31

The Role and Effectiveness of Community Councils with Regard to Community Consultation: Robina Goodlad, John Flint, Ade Kearns, Margaret Keoghan, Ronan Paddison & Mike Raco. (1999) (£5.00)

Perceptions of Local Government: A Report of Focus Group Research: Carole Millar Research. (1999) (£5.00)

Supporting Parenting in Scotland: Sheila Henderson. (1999) (£5.00)
Summary available: Social Work Research Findings No.33

Investigation of Knife Stab Characteristics: I. Biomechanics of Knife Stab Attacks; II. Development of Body Tissue Simulant: Bioengineering Unit & Department of Mechanical Engineering, University of Strathclyde. (1999) (£5.00)

City-Wide Urban Regeneration: Lessons from Good Practice: Professor Michael Carley & Karryn Kirk, School of Planning & Housing, Heriot-Watt University. (1999) (£5.00)
Summary available: Development Department Research Findings No.66

An Examination of Unsuccessful Priority Partnership Area Bids: Peter Taylor, Ivan Turok & Annette Hastings, Department of Urban Studies, University of Glasgow. (1999) (£5.00)
Summary available: Development Department Research Findings No.67

The Community Impact of Traffic Calming Schemes: Ross Silcock Ltd, Social Research Associates. (1999) (£10.00)
Summary available: Development Department Research Findings No.68

The People's Panel in Scotland: Wave 1 (June-September 1998): Dr Nuala Gormley. (1999) (Free)
Summary only available: General Research Findings No.1

The People's Panel in Scotland: Wave 2 (August-November 1998): Dr Nuala Gormley. (1999) (Free)
Summary only available: General Research Findings No.2

Evaluation of Prevention of Environmental Pollution from Agricultural Activity (PEPFAA) Code: Peter Evans, Market Research Scotland. (1999) (£5.00)
Summary available: General Research Findings No.3

Review of Safer Routes to School in Scotland: Derek Halden Consultancy in association with David McGuigan. (1999) (£5.00)

Climate Change: Scottish Implications Scoping Study: Andrew Kerr & Simon Allen, University of Edinburgh; Simon Shackley, UMIST; Ronnie Milne, Institute of Terrestrial Ecology. (1999) (£5.00)
Summary available: Environment Group Research Findings No.5

City-Wide Urban Regeneration: Lessons from Good Practice: Professor Michael Carley & Karryn Kirk. (1999) (£5.00
Summary available: Development Department Research Findings No.66

An Examination of Unsuccessful Priority Partnership Area Bids: Peter Taylor, Ivan Turok & Annette Hastings. (1999) (£5.00)
Summary available: Development Department Research Findings No.67

The Children's Traffic Club in Scotland: Katie Bryan-Brown & Gordon Harland. (1999) (£5.00)
Summary available: Development Department Research Findings No.69

An Evaluation of the New Life for Urban Scotland Initiative in Castlemilk, Ferguslie Park, Wester Hailes and Whitfield: Cambridge Policy Consultants. (1999) (£10.00)
Summary available: Development Department Research Findings No.70

National Monitoring and Interim Evaluation of the Rough Sleepers Initiative in Scotland: Anne Yanetta & Hilary Third (School of Planning & Housing, ECA/Heriot-Watt University) & Isobel Anderson (HPPU, University of Stirling). (1999) (£5.00)
Summary available: Development Department Research Findings No.71

Social Inclusion Research Bulletin No.2. (1999) (Free)

Costs in the Planning Service: Paula Gilder Consulting. (1999) (£5.00
Summary available: Development Department Research Findings No.72

Evaluation of the Teenwise Alcohol Projects: Simon Anderson & Beckie Sawyer. (1999) (£6.00)
Summary available: Crime and Criminal Justice Research Findings No.34

The Work of Precognition Agents in Criminal Cases: David J Christie & Susan R Moody (University of Dundee). (1999) (£5.00)
Summary available: Crime and Criminal Justice Research Findings No.32

Counting the Cost: Crime Against Business in Scotland: John Burrows, Simon Anderson, Joshua Bamfield, Matt Hopkins & Dave Ingram. (1999) (£10.00)
Summaries available: Crime and Criminal Justice Research Findings No's. 35, 38, 39 & 40.

Park and Ride in Scotland: Transport Research Laboratory and Strathclyde Passenger Transport. (1999) (£5.00)
Summary available: Development Department Research Findings No.74

Understanding Offending Among Young People: Janet Jamieson, Gill McIvor & Cathy Murray. (1999) (£16.00)
Summary available: Social Work Research Findings No.37

The View from Arthur's Seat: A Literature Review of Housing and Support Options 'Beyond Scotland': Ken Simons & Debbie Watson (Norah Fry Research Centre, University of Bristol). (1999) (£5.00)

"If You Don't Ask You Don't Get": Review of Services to People with Learning Disabilities: The Views of People who use Services and their Carers: Kirsten Stalker, Liz Cadogan, Margaret Petrie, Chris Jones, Jill Murray (Scottish Human Services). (1999) (£5.00)

Diversion from Prosecution to Social Work and Other Service Agencies: Evaluation of the 100% Funding Pilot Programmes: Monica Barry & Gill McIvor (University of Stirling). (1999) (£5.00)
Summary available: Crime and Criminal Justice Research Findings No.37

Council Tax Collection Arrangements in Scotland, England & Wales: Institute of Revenues, Rating and Valuation. (1999) (£5.00)
Summary available: Development Department Research Findings No.80 (2000)

Why People don't Drive Cars: Sue Granville & Andra Laird (George Street Research). (1999) (£5.00)

Support at Home – Views of Older People about their Needs and Access to Services: Charlotte MacDonald (University of Stirling). (1999) (£14.00)
Summary available: Social Work Research Findings No.35

Transport Provision for Disabled People in Scotland: Sheila Henderson and Brian Henderson, Reid Howie Associates. (1999) (£10.00)
Summary available: Development Department Research Findings No. 76

Community Mediation in Scotland – A Study of Implementation: Robert E Mackay and Amanda J Brown, University of Dundee. (1999) (£5.00)
Summary available: Legal Studies Research Findings No.24

Drug Misuse in Scotland: Simon Anderson & Martin Frischer. (2000) (£5.00)
Summary available: Crime and Criminal Justice Research Findings No.17

Support for Majority and Minority Ethnic Groups at Home – Older People's Perspectives: Alison Bowes and Charlotte MacDonald
Summary only available: Social Work Research Findings No.36

Intermediate Diets, First Diets and Agreement of Evidence in Criminal Cases: An Evaluation: Frazer McCallum & Professor Peter Duff (Aberdeen University Faculty of Law). (2000) (£5.00)
Summary available: Crime and Criminal Justice Research Findings No.42

The Experience of Violence and Harassment of Gay Men in the City of Edinburgh: Colin Morrison & Andrew Mackay (The TASC Agency). (2000) (£5.00)
Summary available: Crime and Criminal Justice Research Findings No.41

The Development of the Scottish Partnership on Domestic Abuse and Recent Work in Scotland: Dr Sheila Henderson (Reid Howie Associates). (2000) (£5.00)

Children, Young People and Crime in Britain and Ireland: From Exclusion to Inclusion - 1998 Conference Papers: Monica Barry (University of Stirling), Joe Connolly (Action for Children), Olwyn Burke, Dr J Curran (Central Research Unit, Scottish Executive). (2000) (£5.00)

Overview of Written Evidence Received as Part of the Review of the Public Health Function in Scotland : Summary available only: General Research Findings No.4

Assessment of the Voter Education Campaign for the Scottish Parliament Elections: (Scotland Office Publication): Andra Laird, Sue Granville & Jo Fawcett (George Street Research). (2000) (£5.00)

Review of the Experience of Community Councils as Statutory Consultees on Planning Applications: Ewan McCraig, MVA. (2000) (£5.00)
Summary available Development Department Research Findings No.77

Family Support and Community Care: A Study of South Asian Older People: Alison Bowes and Naira Dar with the assistance of Archana Srivastava (University of Stirling). (2000) (£6.00)
Summary available: Social Work Research Findings No.38

Review of the Experience of Community Councils as Statutory Consultees on Planning Applications: Ewan McCraig, MVA. (2000) (£5.00)
Summary Available Development Department Research Finding No.77

Development Department Research 2000-2001: (2000) (Free)

An Evaluation of the SACRO (Fife) Young Offender Mediation Project: Becki Sawyer, System 3. (2000) (£5.00)
Summary available: Crime and Criminal Justice Research Findings No.43

Development Department Research 2000-2001: (2000) (Free)

Environment Group Research Programme 2000-2001: (2000) (Free)

Social Inclusion Bulletin No.3: (2000) (Free)

Road Safety in the Scottish Curriculum: Tony Graham, ODS Ltd. (2000) (£5.00)
Summary available: Development Department Research Findings No.78

The Role of Information and Communications Technology in Road Safety Education: BITER – The British Institute of Traffic Education Research. (2000) (£5.00)
Summary available: Development Department Research Findings No.79

Evaluation of Scottish Road Safety Campaign Travel Packs: Sharon Reid, Andra Laird & Jo Fawcett. (2000) (£5.00)
Summary available: Development Department Research Findings No.82

Audit of ICT Initiatives: In Social Inclusion Partnerships and Working for Communities Pathfinders in Scotland: Joanna Gilliatt, Doug Maclean & Jenny Brogden, Lambda Research & Consultancy Ltd. (2000) (£5.00)

Researching Ethnic Minorities in Scotland: Reid-Howie Associates. (2000) (Free)

A Comparative Evaluation of Greenways and Conventional Bus Lanes: Colin Buchanan and Partners. (2000) (£5.00)
Summary available: Development Department Research Findings No.83

Advertising Planning Proposals: James Barr Planning Consultants. (2000) (£5.00)
Summary available: Development Department Research Findings No.84

Developing Markets for Recyclable Materials in Scotland: Prioritising Materials: Enviros RIS Ltd in association with Clean Washington Centre. (2000) (Free).
Summary only available: Environment Group Research Findings No.6

The Development of the Scottish Partnership on Domestic Abuse and recent Work in Scotland: Dr S Henderson, Reid Howie Associates. (2000) (£5.00).

Evaluation of the Airborne Initiative (Scotland): Gill McIvor, Vernon Gayle, Kirstina Moodie, Stirling University and Ann Netten, University of Kent. (2000) (£5.00)
Summary available: Crime and Criminal Justice Research Finding No.45

A Review of the Research Literature on Serious and Sexual Offenders: Clare Connelly and Shanti Williamson, University of Glasgow. (2000) (£8.00).

Summary available: Crime and Criminal Justice Research Finding No.46

The Quality of Services in Rural Scotland: Steven Hope, Simon Anderson and Becki Sawyer, System Three.
Summary available: Rural Affairs Research Finding No.5.

Social Exclusion in Rural Areas; A Literature Review and Conceptual Framework: Mark Shuckshank and Lorna Philip, University of Aberdeen. (2000) (£10).
Summary available: Rural Affairs Research Finding No.6.

Charities Report: **1. Scottish Charity Legislation; Full Report** (2000) (£15.00).
2. Scottish Charity Legislation; Executive Summary (2000) (Free).
3. Scottish Charity Legislation; Annexes (2000) (£5.00).
4. Public Charitable Collections (2000) ((£5.00).
5. Public Trusts and Educational Endowments (2000) (£5.00).
University of Dundee.
Summary available: Legal Studies Research Finding 26

Meeting in the Middle: A Study of Solicitors' and Mediators Divorce Practice: Fiona Myers and Fran Wasoff. University of Edinburgh. (2000) (£5.00)
Summary available: Legal Studies Research Finding No.25.

Real Burdens: Survey of Owner Occupiers' Understanding of Title Conditions: Andra Laird and Emma Peden, George Street Research (2000) (£5.00)
Summary available: Legal Studies Research Finding No.27

Survey of Complainers to the Scottish Legal Services Ombudsman: The Customer Management Consultancy Ltd (2000) (£5.00)

An Evaluation of Electronically Monitored Restriction of Liberty Orders: david Lobley and David Smith, Lancaster University (2000) (£5.00)
Summary available: Crime and Criminal Justice Research Finding No.47

Interviewing and Drug Testing of Arrestees in Scotland: A Pilot Study of the Arrestee Drug Abuse Monitoring (ADAM) Methodology: Neil McKeganey, Clare Connelly, Lesley Reid & John Norrie University of Glasgow, Janusz Knepil Gartnavel General Hospital Glasgow. (2000) (£5.00)
Summary available: Crime and Criminal Justice Research Finding No 48

The Role of Sport in Regenerating Deprived Areas: Fred Coalter with Mary Allison and John Taylor, Centre for Lesisure Research (2000) (£5.00)
Summary available: Development Department Research Finding No 86

The Role of Pre-Application Discussions and Guidance in Planning: Peter Gibson and Robert Stevenson, The Customer Management Consultancy Ltd (2000) (£5.00)
Summary available: Development Department Research Finding No 85

"Huts and Hutters" in Scotland: Research Consultancy Services (2000) (£5.00)

Research Finding Only: Motivations to Public Service: Development Department RF No. 87. Sue Granville and Andra Laird, George Street Research (2000) (£0.00)

Research Finding Only: The What, Where and When of Being a Councillor: Development Department RF No. 88. Paolo Vestri and Stephen Fitzpatrick, Scottish Local Government Information Unit. (2000) (£0.00)

Further information on any of the above is available by contacting:

Dr A Scott
Chief Research Officer
The Scottish Office Central Research Unit
Room J1-5
Saughton House
Broomhouse Drive
Edinburgh
EH11 3XA

or by accessing the World Wide Website: www.scotland.gov.uk